CONTENTS

Words in the glossary appear in **bold** the first time they are used in the text.

New Language Loading...

Have you ever tried to learn a new language? It can seem hard at first, but the more you practice, the closer you get to being able to talk to another person in that language. Learning a **programming language** can be a similar process, except you're learning how to talk to a computer instead of a person!

There are many programming languages out there, and they each have their own features and uses. Programmers can use the Ruby language for many things. It's easy to use, making it a great coding language to learn if you're a beginner.

< GET CODING! >

Get Coding with Ruby™

By Jenna Vale

Gareth Stevens
PUBLISHING

Please visit our website, www.garethstevens.com. For a free color catalog of all our high-quality books, call toll free 1-800-542-2595 or fax 1-877-542-2596.

Library of Congress Cataloging-in-Publication Data
Names: Vale, Jenna, author.
Title: Get coding with Ruby / Jenna Vale.
Description: Buffalo, New York : Gareth Stevens Publishing, [2024] | Series: Get coding! | Includes bibliographical references and index.
Identifiers: LCCN 2023014700 | ISBN 9781538288702 (library binding) | ISBN 9781538288696 (paperback) | ISBN 9781538288719 (ebook)
Subjects: LCSH: Ruby (Computer program language)–Juvenile literature.
Classification: LCC QA76.73.R83 .V349 2024 | DDC 005.13/3–dc23/eng/20230414
LC record available at https://lccn.loc.gov/2023014700

Published in 2024 by
Gareth Stevens Publishing
2544 Clinton Street
Buffalo, NY 14224

Portions of this work were originally authored by Patricia Harris and published as *Understanding Coding with Ruby*. All new material in this edition authored by Jenna Vale.

Designer: Leslie Taylor
Editor: Megan Kellerman

Photo credits: Cover (main image) TippaPatt/Shutterstock.com, (illustrations) StonePictures/Shuttertstock.com; cover & series art (coding background) BEST-BACKGROUNDS/Shutterstock.com; p. 5 Mathias Meyer/https://commons.wikimedia.org/wiki/File:Yukihiro_Matsumoto_EuRuKo_2011.jpg; p. 7 Melisa Parlak/Shutterstock.com; p. 9 Ground Picture/Shutterstock.com; p. 11 Kostsov/Shutterstock.com; pp. 13, 15 (Ruby screenshots) courtesy of www.ruby-lang.org; p. 15 ESB Professional/Shutterstock.com; p. 17 Prostock-studio/Shutterstock.com; p. 19 carballo/Shutterstock.com; p. 21 LightField Studios/Shutterstock.com; p. 23 Prostock-studio/Shutterstock.com; p. 25 Monkey Business Images/Shutterstock.com; p. 29 Tyler Olson/Shutterstock.com.

Printed in the United States of America

CPSIA compliance information: Batch #CS24GS: For further information contact Gareth Stevens at 1-800-542-2595.

This is Yukihiro Matsumoto, a Japanese computer scientist and the creator of Ruby.

Ruby's Creator

Ruby is open-source **software**, which means it's a free computer program that can be used and even changed by anyone. When Yukihiro Matsumoto originally created Ruby, his goal was to combine features of other programming languages to make a language that felt natural to use.

Writing Code with Words

You can write code with programming languages that use either images or words. Ruby is a standard text-based, or word-based, programming language. It uses lines of text to give a computer instructions to be **executed** in Ruby. Because of this, Ruby users need to know a set of special words called commands. These words tell the program how to reply, or they give users important data.

Ruby is a powerful programming language. It offers users a lot of choices. It's also a step closer to using harder written-code languages, such as C++, Perl, and Java.

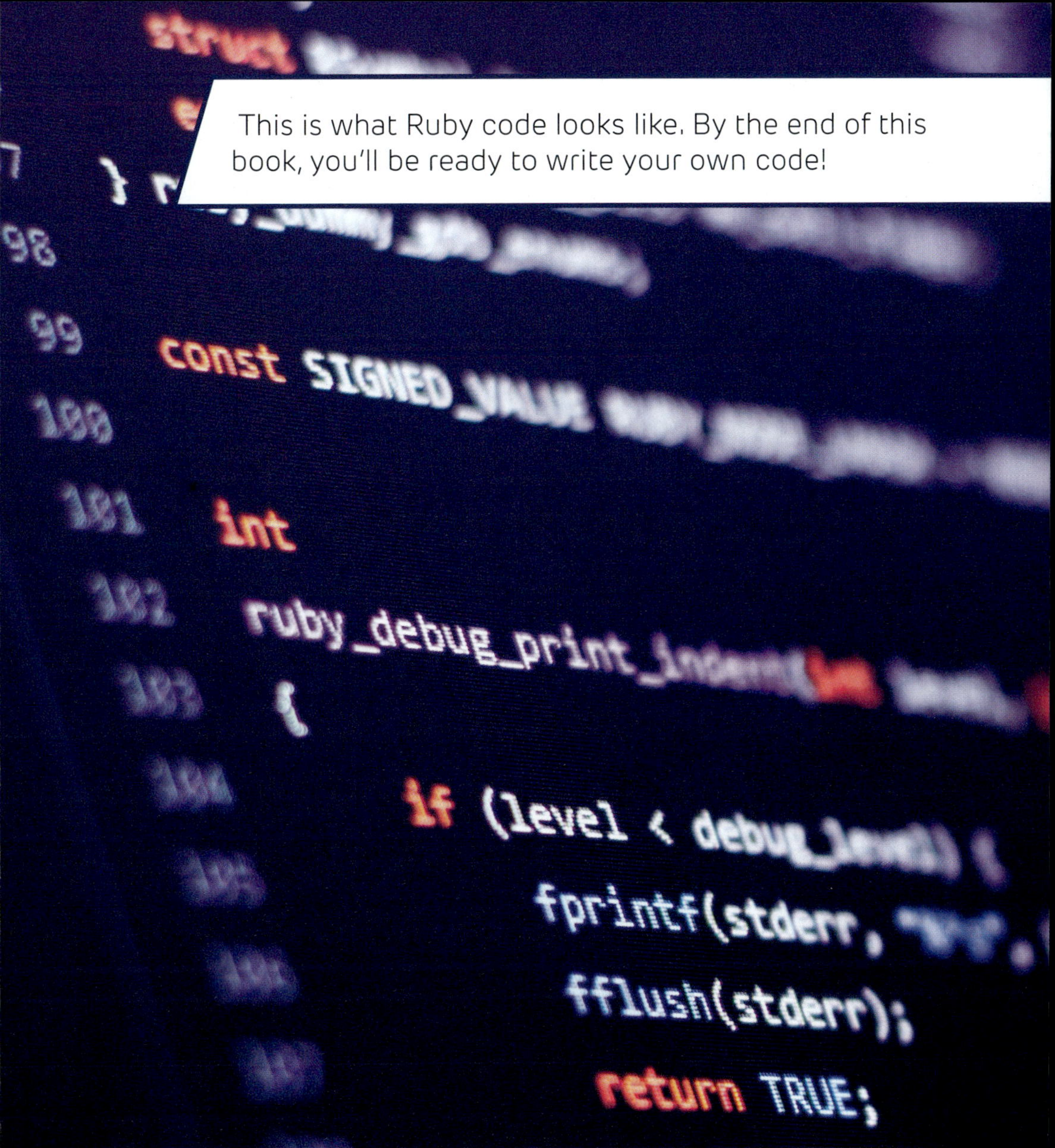

This is what Ruby code looks like. By the end of this book, you'll be ready to write your own code!

Coding with Images

Unlike text-based Ruby, some programming languages use image blocks that coders drag and drop into a workspace to write code. The blocks have text commands built into them. This coding method uses a **graphical** user interface to allow users to communicate with the computer. This kind of coding can be done with programs such as Scratch or Hopscotch.

Following the Rules

Coding in any language is about following rules. It's almost like playing a game, and it can be just as much fun.

Rule 1: Coders must know what they want the computer to do and write a plan.

Rule 2: Coders must use special words to have the computer take input, make choices, and act.

Rule 3: Coders need to think about what tasks can be put into a group.

Rule 4: Coders need to use **logic** with AND, OR, NOT, and other logic statements as key words.

Rule 5: Coders must explore the **environment** and understand how it works.

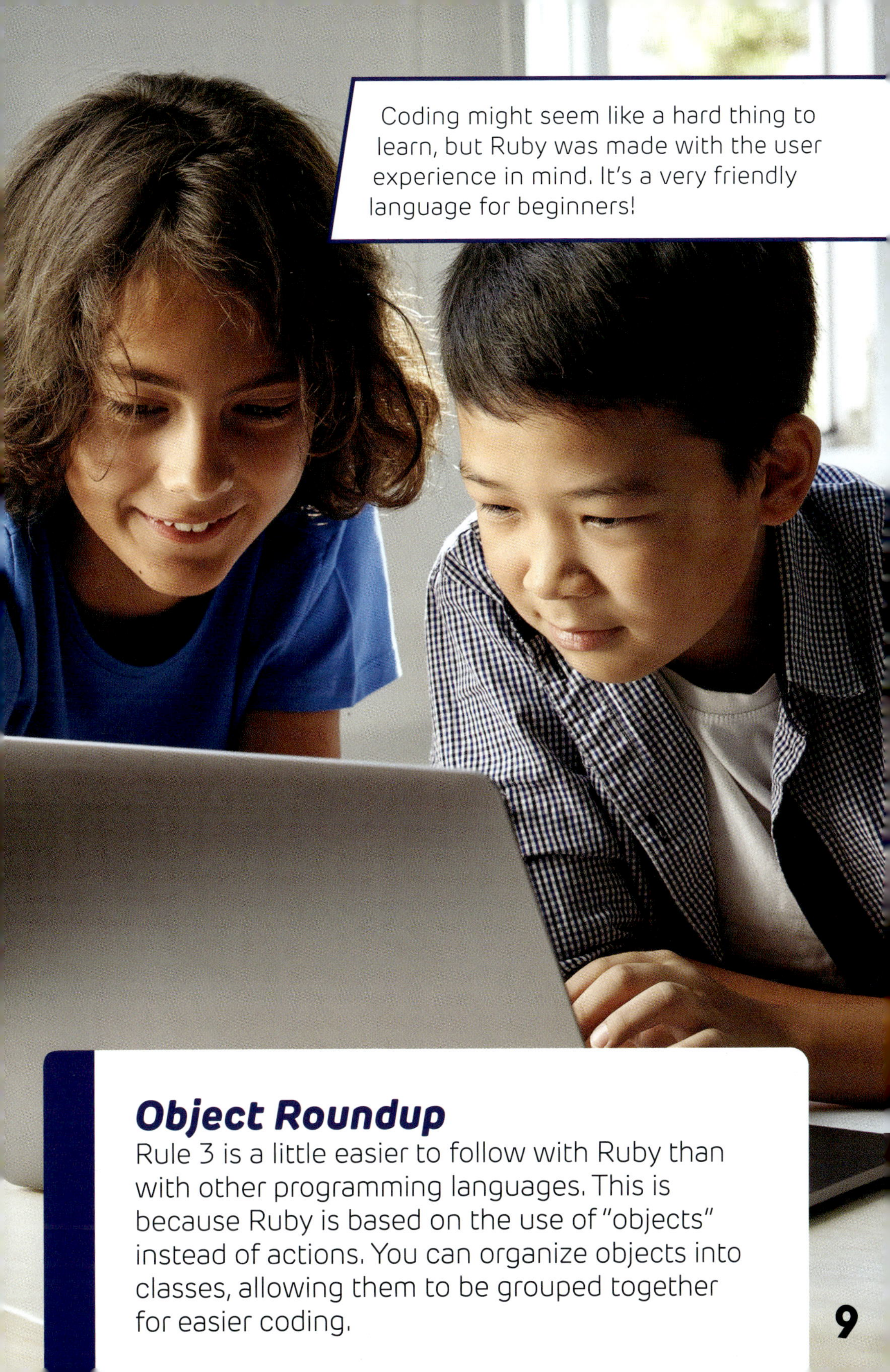

Coding might seem like a hard thing to learn, but Ruby was made with the user experience in mind. It's a very friendly language for beginners!

Object Roundup

Rule 3 is a little easier to follow with Ruby than with other programming languages. This is because Ruby is based on the use of "objects" instead of actions. You can organize objects into classes, allowing them to be grouped together for easier coding.

Show Your Work

When coding with Ruby, you'll use English words and special characters that mean specific things. Sometimes the words mean just what you expect, but sometimes they have different meanings.

The command "print" displays the results of the code you typed. If you write "print 5," the computer displays "5" on your screen. The command "print 2 + 5" displays "7." The math work is built in.

The "puts" command does the same thing but uses more than one line. If you write "puts 2 + 5," your screen shows "7" and then another line beneath it. The second line will be blank or display "=>nil" depending on where you're writing your Ruby code.

These are some of the commands you'll use when coding with Ruby.

print/puts — Displays the results of a program on the screen.

#, =begin, =end — Tells Ruby the text after # (on a single line) or between =begin and =end (on multiple lines) is a comment that's not part of the code.

chomp — Tells Ruby to ignore line breaks, and displays user input on a single line.

gets — Tells Ruby to collect user input, such as the user's name.

if/else — Allows Ruby to make a decision based on input.

elsif — Used when user input could result in multiple answers.

nil — An object with no value. It's also the Ruby program's way of saying "I'm done."

Seeing Is Believing

If you're practicing in a free web-based editor, the "puts" command will display a blank line beneath your code in the output window. To see that the "puts" command adds a second line, practice using the math problem from the last page. If you write "puts 2 + 5" and then "print 2 + 5," you'll get 7 listed on two separate lines in the output window.

The Editor and Terminal

To begin working with Ruby, you have a few choices. You can do a beginner's lesson or practice writing code at www.try.ruby-lang.org. This is a free text editor and output window on the official website for Ruby. It will let you write code and test how it runs right away!

You can also use a text editor that your computer already has, or ask an adult for help **downloading** a free one from the internet. The text editor allows you to write code, but you need a **terminal** window to show the output of the program.

This is one example of what a text editor with working Ruby code looks like.

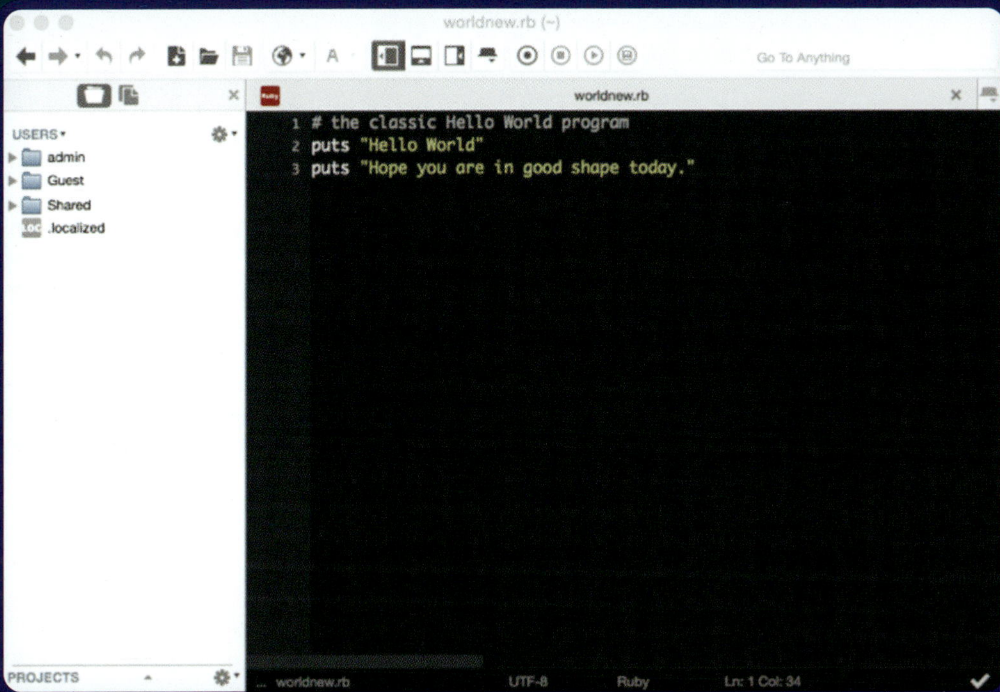

Find the Perfect Fit

If you're unsure which editor or terminal to try, test out the editor and terminal on Ruby's website. You can also try another web-based editor such as JDoodle or OneCompiler. If you want to go ahead and write programs you can save to your computer right away, check out the free Komodo IDE text editor.

If you're working with a text editor on your computer, save the program when you're ready to test it. Name the file and change the extension to .rb so Ruby will know it's a Ruby file. Once you've written a program and saved it as a Ruby file, open Ruby to run the program. On Apple computers, you need to open a terminal window and type in "irb" to open Ruby. On PCs, the Ruby program appears under the Start menu.

Now that you know how saving and running a Ruby program works, you're ready to try your first coding project.

Seeing Nil

Remember the "nil" object? If you practice in a web-based editor, you might not see "nil" come up! But if you're working in an editor and terminal you've downloaded to your computer, "nil" will show up at the end of your program. It's Ruby's way of saying "I'm done, and I've got nothing to add."

This is the terminal screen on an Apple computer. You can use it to run Ruby code once you've saved your file in a text editor.

Simple Plan, Simple Program

No matter what language you're programming in, the first program you'll learn is usually the Hello World program. The plan for this program is to have the computer display the words "Hello World." This is a very simple plan, and the code is also very simple.

Hello World
a simple program to print out Hello World
puts "Hello World"

The first line starts with #, which is called a hash. Use that character when adding a comment to your code. Comments are helpful for keeping track of what each part of your program is supposed to do. They don't affect how the program runs.

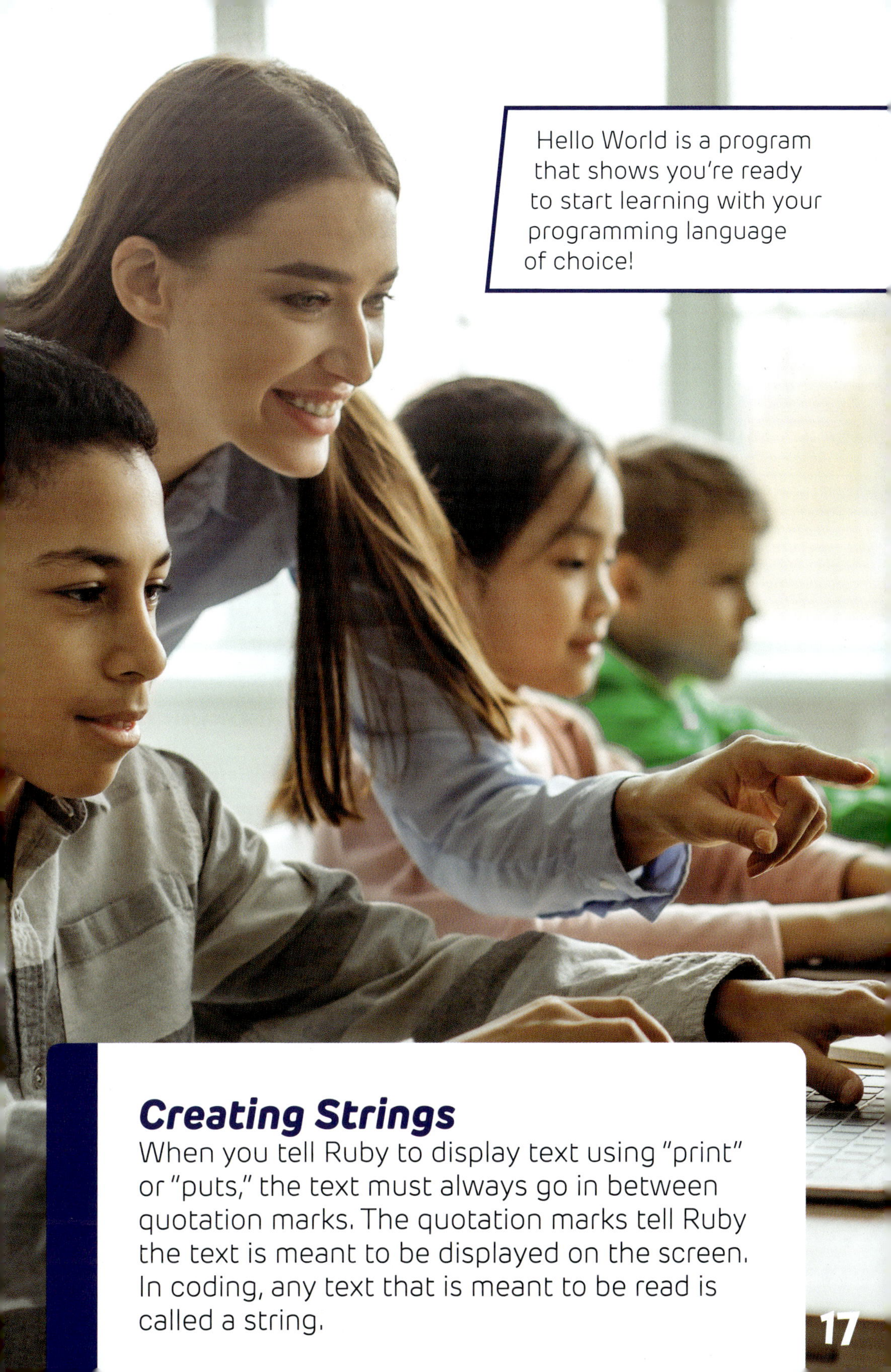

Hello World is a program that shows you're ready to start learning with your programming language of choice!

Creating Strings

When you tell Ruby to display text using "print" or "puts," the text must always go in between quotation marks. The quotation marks tell Ruby the text is meant to be displayed on the screen. In coding, any text that is meant to be read is called a string.

Talking with Ruby

Sometimes you might need your program to talk to the user. In this case, the program would ask the user for input, wait for the user to provide input, and then continue running. A program might need user input because it's gathering data or because it needs a user to make a choice about something.

If you want to ask a user for their name and have Ruby greet them, the code would look like this:

Greeting the User

```
puts "Hello User."
puts "Please tell me your name."
name = gets.chomp
puts "#{name}, I hope you are having fun today!"
```

Adding "#{name}" to the string is what allows Ruby to add the user's name into the greeting. This is called string interpolation.

Defining Variables

Did you notice the equal sign in line 3 of the code? This is how you identify a **variable** in Ruby. You choose a label for the variable ("name") and then tell Ruby what the variable is. In this case, the variable is whatever the user input is. Ruby reads user input through "gets."

Add to the code if you want more input from the user or if you want Ruby to handle the input a certain way. For example, if you want Ruby to capitalize the user's name even if the user enters all lowercase letters, you can add a new line to the code:

Capitalizing the User's Name	
3	name = gets.chomp
4	name.capitalize!
5	puts "#{name}, I hope you are having fun today!"

Line 4 now tells Ruby to capitalize the user's name when it prints the next line. The dot before "capitalize" means you are about to use a method in Ruby—more on methods in the next chapter!

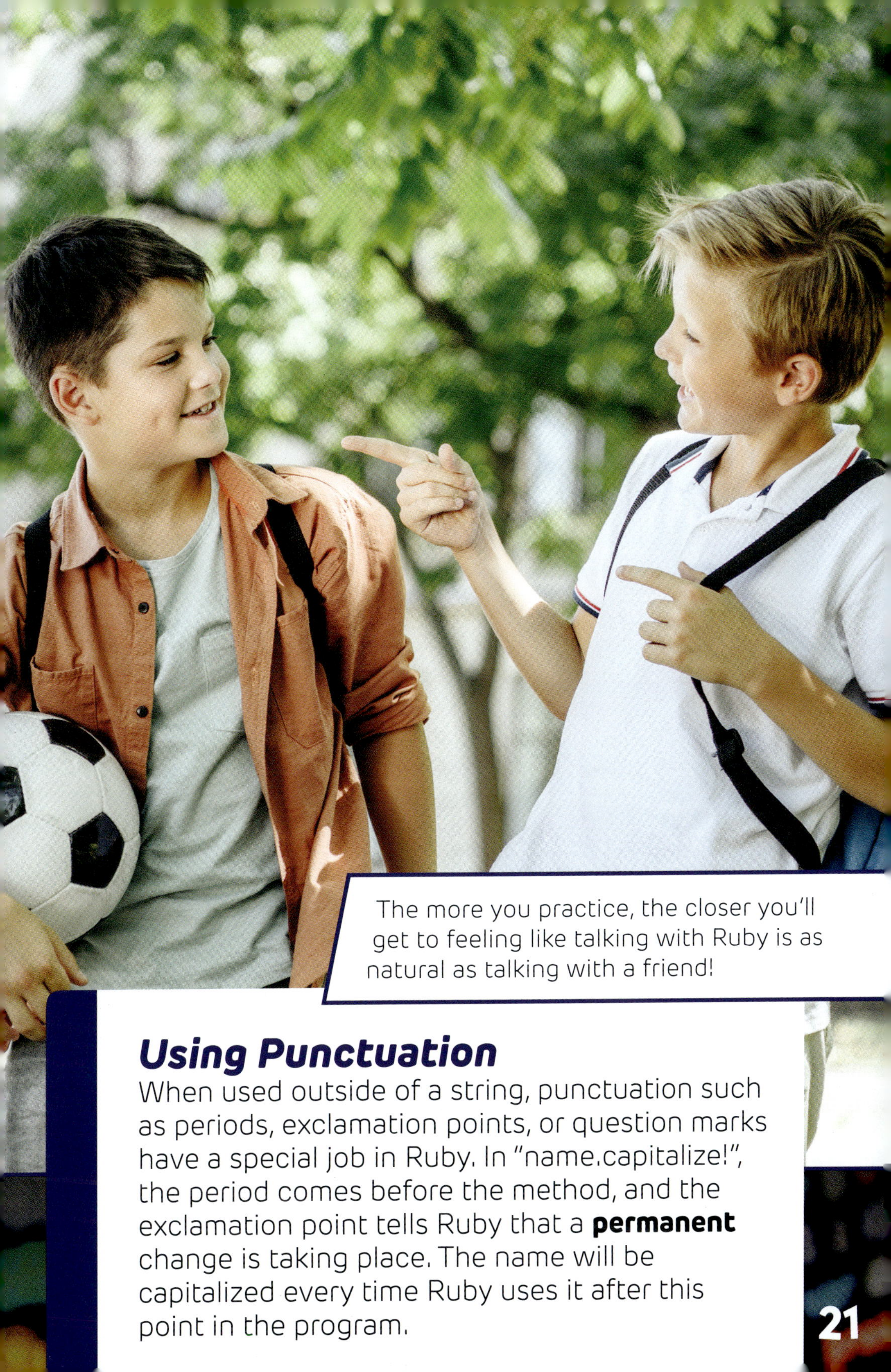

The more you practice, the closer you'll get to feeling like talking with Ruby is as natural as talking with a friend!

Using Punctuation

When used outside of a string, punctuation such as periods, exclamation points, or question marks have a special job in Ruby. In "name.capitalize!", the period comes before the method, and the exclamation point tells Ruby that a **permanent** change is taking place. The name will be capitalized every time Ruby uses it after this point in the program.

Ruby's Methods

In Ruby, a method is a set of premade instructions. Ruby has some methods, such as "capitalize," that are built in. You can also make your own methods! This is called declaring a method, and you use "def" at the beginning of a line of code to define the method. It's like giving value to a variable. When you're done defining the method, add "end" on a new line to close it off.

Adding a period after a variable tells Ruby that you want to use those instructions. This is how you call a method into action.

Don't forget that writing any code, including a new Ruby method, requires a plan. Take notes or write comments with the # character to keep track of everything!

Metaprogramming

Ruby has some methods built in, and you can create methods yourself. Methods also happen a third way: Ruby can write methods by itself! This is called metaprogramming. It happens when Ruby has enough data to create methods based on what you've already defined for it.

Ruby has other methods you can practice with too. If you print text with the reverse method, the text will come out backward! For example, the code **print "Hello World".reverse** will display **dlroW olleH** on the screen.

You can also use methods one after the other in one line of code. The order in which you use them matters! See the examples below.

Code: **print "jessie".capitalize.reverse**
Output: **eisseJ**
Code: **print "jessie".reverse.capitalize**
Output: **Eissej**

Do you see the difference? Ruby will use the methods in the same order you place them in while coding. This is called method chaining.

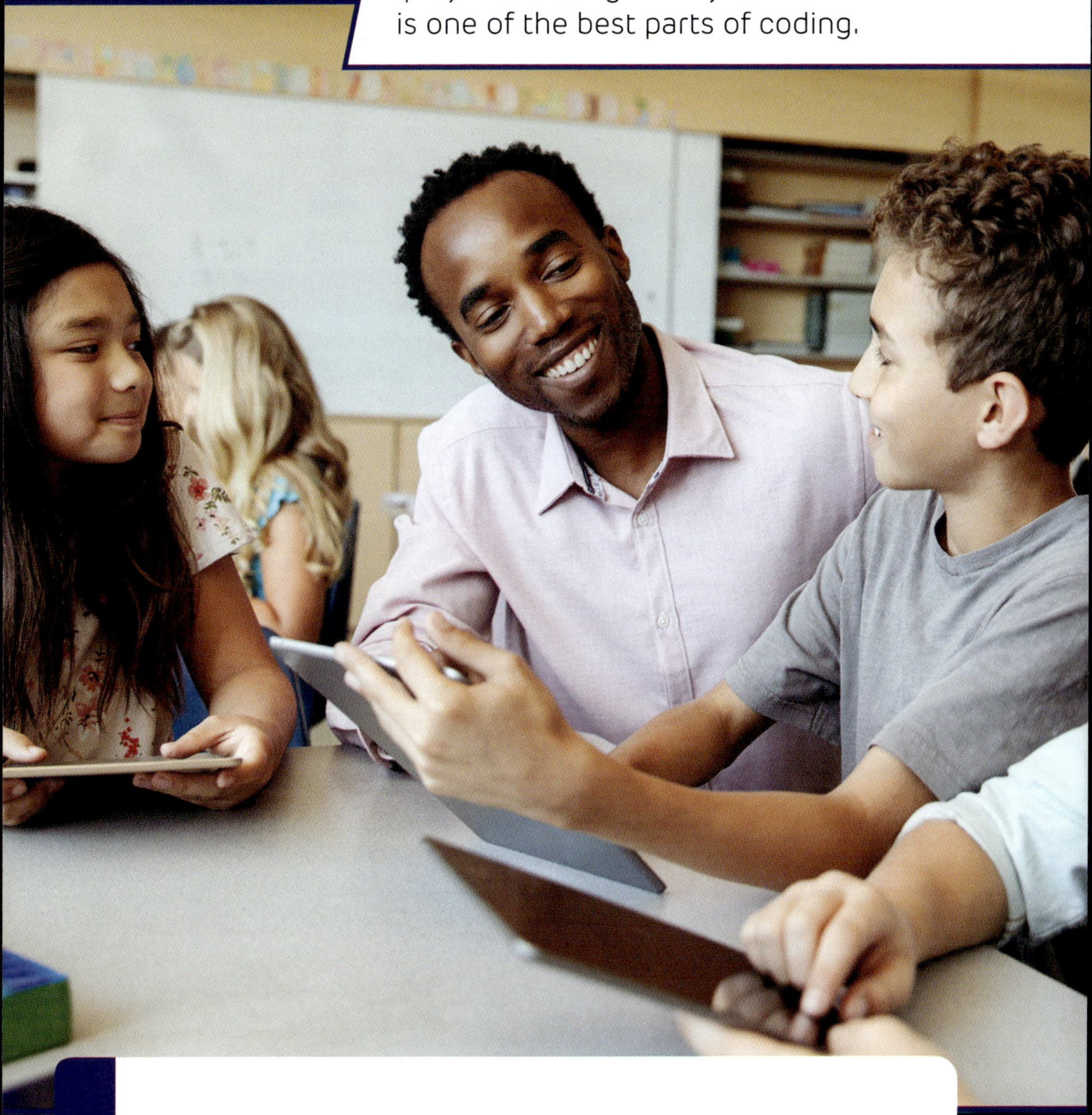

Coders often work together when creating a project. Sharing what you learn with others is one of the best parts of coding.

Exploring Further

Methods can do more than change one word in your code. If you save a string of text to a variable, you can also use certain methods to change the text around. For example, you can use the "gsub", or global substitute, method to change one word to another in the string.

Test Your Skills

Now that you've learned some basic commands and methods, the next step is to add logic. The commands "if" and "else" can help you make things more interesting! They allow the program to make a choice based on user input.

A similar command is called "elsif". This command word is made up of "else" and "if" put together. This command tells Ruby what to do if the user provides one of many possible answers. When writing the code, start with "if" and then use "elsif" instead of the "else" command. See the example on the next page to learn more.

This code starts at line 6 because it adds to the program we were writing on page 20. What other questions would you want Ruby to ask the user?

Adding Logic		
6	puts "#{name}, which color do you like?"	Ruby will address the user with the name they gave earlier and ask for an answer. The answer is defined as what the user enters and will be displayed on a single line.
7	puts "Choose one of the following:"	
8	puts "red, blue, or green."	
9	answer = gets.chomp	
10	answer.downcase!	Ruby will use the downcase method to make sure the user's answer is all lowercase.
11	# using elsif for choices	This is how you use "elsif" logic. This tells Ruby how to reply if the user's answer is red, blue, or green.
12	if "red" == answer	
13	puts "AH HA! Your choice is red."	
14	elsif "blue" == answer	
15	puts "AH HA! Your choice is blue."	
16	elsif "green" == answer	
17	puts "AH HA! Your choice is green."	
18	end	

Logic Operators

Working with commands like if, else, and elsif creates certain conditions that must be met for Ruby to continue running the program. You can add more conditions for Ruby to work in by using logic operators too. A few logic operators include "&&" (meaning AND), "||" (meaning OR), and "not".

Create with Ruby!

Ruby is one of the most popular programming languages used for creating web **applications**. It's special because it works with objects instead of actions, but it is also a general-purpose language. This means it can be used to write many different types of programs.

Learning to code with Ruby is free and easy, no matter what your coding experience is. The more you learn, the more you can create! You can code games with Ruby 2D, create web platforms with Ruby on Rails, and even gain experience to learn more coding languages in the future. It's time to get started!

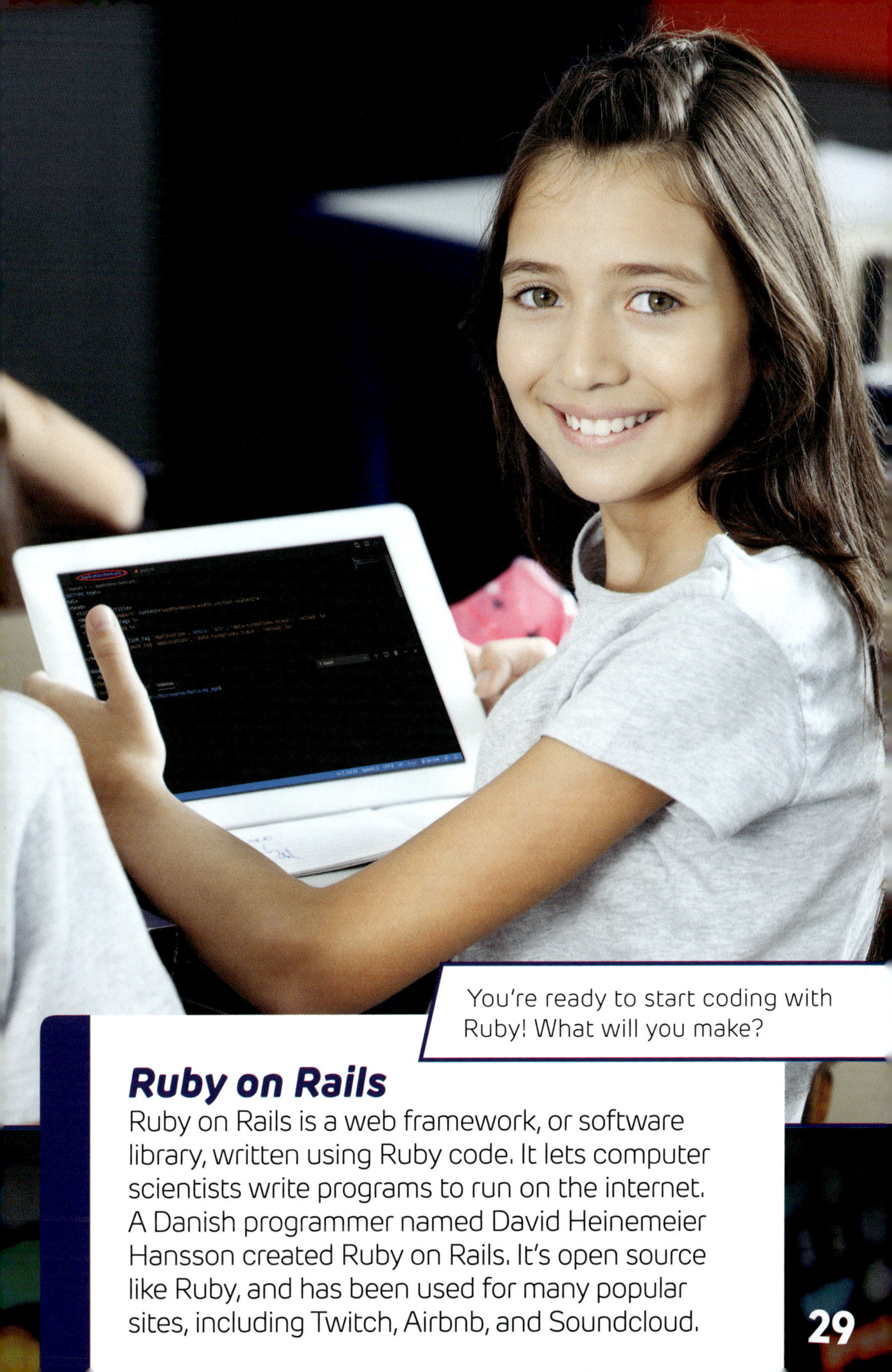

You're ready to start coding with Ruby! What will you make?

Ruby on Rails

Ruby on Rails is a web framework, or software library, written using Ruby code. It lets computer scientists write programs to run on the internet. A Danish programmer named David Heinemeier Hansson created Ruby on Rails. It's open source like Ruby, and has been used for many popular sites, including Twitch, Airbnb, and Soundcloud.

GLOSSARY

application: A computer program that serves a specific purpose.

download: To move or copy a file or program from one computer or device to another.

environment: The combination of computer hardware and software that allows a user to perform various tasks.

execute: To carry out or put into effect, such as a plan.

graphical: Having to do with graphics, or pictures.

logic: A proper or reasonable way of thinking about or understanding something.

permanent: Unable to be reversed.

programming language: A coding language designed to give instructions to a computer.

software: A program that runs on a computer and performs certain tasks.

terminal: A computer screen or window where computer data can be viewed.

variable: Something that can be changed.

FOR MORE INFORMATION

Books

González, Echo Elise. *Logic in Coding*. Chicago, IL: World Book, 2020.

Lynn, Randy. *Coding Concepts for Kids*. Emeryville, CA: Rockridge Press, 2020.

McCue, Camille. *Getting Started with Coding: Get Creative with Code!* Hoboken, NJ: Wiley, 2020.

Websites

Code.org
www.code.org
An educational nonprofit focused on providing computer science tools for grades K–12.

Ruby
www.ruby-lang.org
This is the official website for Ruby. You can download Ruby and learn more about how to code here.

Ruby on Rails
www.rubyonrails.org
This is the official website for Ruby on Rails. Learn more about creating web platforms here.

INDEX